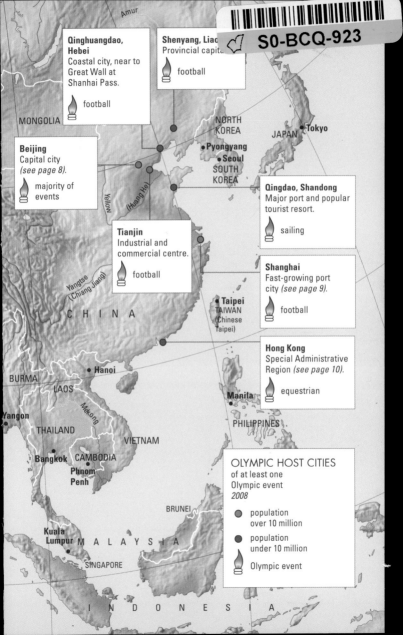

Qinghuangdao, Hebei
Coastal city, near to Great Wall at Shanhai Pass.

football

Shenyang, Lia...
Provincial capita...

football

S0-BCQ-923

Beijing
Capital city
(see page 8).

majority of events

Qingdao, Shandong
Major port and popular tourist resort.

sailing

Tianjin
Industrial and commercial centre.

football

Shanghai
Fast-growing port city *(see page 9)*.

football

Hong Kong
Special Administrative Region *(see page 10)*.

equestrian

OLYMPIC HOST CITIES
of at least one Olympic event
2008

⬤ population over 10 million

● population under 10 million

Olympic event

MONGOLIA

NORTH KOREA

SOUTH KOREA

JAPAN

Tokyo

Pyongyang

Seoul

Amur

Yellow

(Huang He)

Yangtse (Chiang Jiang)

C H I N A

Taipei
TAIWAN
(Chinese Taipei)

Hanoi

BURMA

LAOS

Mekong

Yangon

THAILAND

VIETNAM

Manila

PHILIPPINES

Bangkok

CAMBODIA

Phnom Penh

BRUNEI

Kuala Lumpur

M A L A Y S I A

SINGAPORE

I N D O N E S I A

Pocket China Atlas

Maps and Facts at Your Fingertips

Stephanie Hemelryk Donald
and Robert Benewick

UNIVERSITY OF CALIFORNIA PRESS

Berkeley   Los Angeles

University of California Press, one of the most distinguished university presses in the United States, enriches lives around the world by advancing scholarship in the humanities, social sciences, and natural sciences. Its activities are supported by the UC Press Foundation and by philanthropic contributions from individuals and institutions. For more information, visit www.ucpress.edu.

University of California Press
Berkeley and Los Angeles, California

Cataloging-in-publication data for this title
is on file with the Library of Congress.

ISBN 978-0-520-25468-8

Produced for the University of California Press by
Myriad Editions
59 Lansdowne Place
Brighton, BN3 1FL, UK
www.MyriadEditions.com

Printed on paper produced from sustainable sources.
Printed and bound in Hong Kong through Phoenix Offset
under the supervision of Bob Cassels,
The Hanway Press, London.

15   14   13   12   11   10   09   08
10   9   8   7   6   5   4   3   2   1

CONTENTS 中

Part One
INTRODUCING CHINA

YIMEI IS 20 YEARS OLD. She was born and raised in Sichuan, where her father worked as a Communist Party official and her mother was a high-school teacher. This conventional intellectual background prepared the family for a settled provincial existence, but not for the demands that Chinese modernity now places on young Chinese. Yimei has recently left her hometown to pursue a university degree at China's top economics university, and is planning a career in finance. She is committed to China's future, and believes that educational achievement will establish her right to contribute to the future of the nation.

Meanwhile, Meima, a young woman of the same age in the Three Gorges area of Sichuan, is also looking for a way out. Her best hope is to be taken on as a maid in a middle-class household in Shanghai or Guangzhou.

With such disparities of opportunity and expectation at play in the new generation, introducing "China" is an almost impossible task. One of the great regional powers in the world, China encapsulates the hyperbole of modern market economics, authoritarianism and a monopolistic state ideology, and a new approach to capitalism that embraces wealth creation but eschews liberal democracy. China is powerful in its scope, scale, talent, and ambition, but weakened by population pressures, its stumbling ecology, and its reliance on political rhetoric to manage a complex social fabric, which includes the lives of these two young women.

Both China and the USA need to face up to the challenges of demographic changes, economic growth, military power and, perhaps most of all, environmental protection. The 21st century may well be dominated by the need for these two economic giants to develop a balanced relationship, based on co-operation.

COMPARISON OF VITAL STATISTICS
2005 or latest available data

China

USA

CHINA

9.6 million sq km (3.7 million sq miles)	land area
1,307.6 million	population
137 people per sq km (354 per sq mile)	
$2,264 bn Gross National Income	
$1,740 per person	
$762 bn annual exports	
$700 bn annual imports	
4,732 million tonnes annual CO$_2$ emissions	
4 tonnes per person	
$35 bn military spend	
3 million (including reserves) military personnel	
21 million registered passenger vehicles	
16 per 1,000 people	

CHINA AND THE USA 中

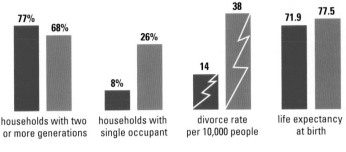

77% | **68%**
households with two
or more generations

8% | **26%**
households with
single occupant

14 | **38**
divorce rate
per 10,000 people

71.9 | **77.5**
life expectancy
at birth

USA

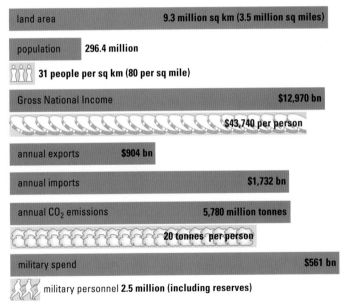

| land area | **9.3 million sq km (3.5 million sq miles)** |

| population | **296.4 million** |

31 people per sq km (80 per sq mile)

| Gross National Income | **$12,970 bn** |

$43,740 per person

| annual exports | **$904 bn** |

| annual imports | **$1,732 bn** |

| annual CO_2 emissions | **5,780 million tonnes** |

20 tonnes per person

| military spend | **$561 bn** |

military personnel **2.5 million (including reserves)**

| registered passenger vehicles | **135 million** |

455 per 1,000 people

As China's political capital, Beijing is home to all central government institutions, and is the powerhouse of China's "knowledge economy". Shanghai is a directly governed municipality, and China's showcase modern city. It is fast turning itself into Asia's telecommunications, logistics and financial hub.

COMPARISON OF VITAL STATISTICS
2006 or latest available data

Beijing

Shanghai

BEIJING

864 completed skyscrapers

82 universities

34 museums

25 public libraries

15.8 million   population

3.8 million migrants

Only 60% of Beijing's *hutongs* (courtyard communities), have survived the developers, and only 430 (33%) of these retain their original character.

772 billion yuan   GDP

industry 30% | **service 69%**   share of GDP

49,505 yuan GDP per person

19,978 yuan disposable income (urban citizens)

640 yuan monthly minimum wage

20 sq meters living area per person

2.97 million   passenger vehicles

66% of days judged clean (Grade II+)   air quality

90% treated   sewage

BEIJING AND SHANGHAI 中

Shanghai will be the "City of Harmony" for the EXPO 2010. Beijing will become the "smiling city" – at least during August 2008.

From 1998 to 2002, 15 square kilometers of Shanghai's old neighbourhoods were demolished. In 2006 alone, 70,000 people were relocated to make way for new developments.

SHANGHAI

completed skyscrapers	**954**

universities **60**

museums **25**

public libraries **28**

population	**18.2 million**

 4.3 million migrants

GDP	**1,029 billion yuan**

share of GDP	**service 51%**	**industry 49%**

 GDP per person **57,310 yuan**

 disposable income (urban citizens) **20,668 yuan**

monthly minimum wage **750 yuan**

 living area per person **16 sq meters**

passenger vehicles **1.1 million**

air quality	**89% of days judged clean (Grade II+)**

sewage	**71% treated**

The most competitive and affluent city in China, Hong Kong has the status of a Special Administrative Region (SAR). Formerly a British colony, it reverted to China in 1997. The "one country, two systems" arrangement allows Hong Kong's 6.9 million residents to retain their social and economic organization, while being subordinated to China's political and territorial sovereignty.

COMPARISON OF VITAL STATISTICS
2005 or latest available data

▇ Mainland China
▨ Hong Kong

MAINLAND CHINA

18,308 billion yuan — GDP

40% services as share of GDP

14,002 yuan GDP per person

16 per 1,000 people number of private passenger vehicles

10 yuan value of energy consumed per 10,000 yuan of GDP

306 cubic meters volume of water consumed per 10,000 yuan of GDP

69.6 life expectancy (men)

73.3 life expectancy (women)

18,364 yuan average annual wage

23 sq meters average living floor space per person

HONG KONG 中

In 2005, only 6 percent of Hong Kong's recovered solid waste was recycled locally. Over 94 per cent was exported, almost all of it to Mainland China.

In 2005, Hong Kong admitted 21,000 skilled workers, and 223,000 unskilled women to work as domestic servants.

61% of Hong Kong's children and adults use the internet.

24% of Hong Kong's residents have a tertiary education.

HONG KONG

GDP **1,593 billion yuan (HK$1,620 billion)**

services as share of GDP **91%**

GDP per person **229,815 yuan (HK$233,565)**

number of private passenger vehicles **52 per 1,000 people**

value of energy consumed per 10,000 yuan of GDP **148 yuan**

volume of water consumed per 10,000 yuan of GDP **1,514 cubic meters**

life expectancy (men) **78.8**

life expectancy (women) **84.4**

average annual wage **118,800 yuan**

average living floor space per person **12 sq meters**

Part Two
CHINA'S PEOPLE

THE SIZE OF CHINA'S OVERALL population is a major factor in the nation's global relevance. Sheer numbers should not mask the diversity of China's people, however. Provincial characteristics and histories, minority nationalities and the deep splits between rural and urban inhabitants all contribute to the complexity of being Chinese. The provinces have distinct characteristics, which are manifested in all aspects of culture, including religion, arts and cuisine. Some of these differences are likened to brand characteristics (Chengdu women are delicate, Shanghai women are clever business women, Shandong men make the best generals and so on) or "local specialities" (Shanxi makes the best vinegar, Sichuan chillies are the hottest and spiciest in China).

Other characteristics are more profound and challenging to the Party State. Minority nationalities, although only 8 percent of the total population, number nearly 100 million people, and include those who are markedly different from the dominant Han majority in religious beliefs and historical allegiances. This is especially so in the Western region, where there are numerous minority populations living in close proximity. The Uighurs, the Mongolians and the Tibetans are just a few of the larger groups scattered across the western provinces, for whom the People's Republic of China is a daily reality but not one that necessarily encapsulates their identities or their hopes. With such diversity in a time of dynamic change, China needs to encourage its people to think of themselves as citizens of a cosmopolitan nation.

Over one-fifth of people in the world –
1.3 billion – live in China. Most people are
Han, but 8 percent belong to one of 55
recognized minority nationalities. The
unnatural predominance of boy babies
throughout the country has created a
gender imbalance which the government
is trying to address.

China's population
would be nearly
400 million higher
without the
government's
family-planning
policy, introduced
in 1979.

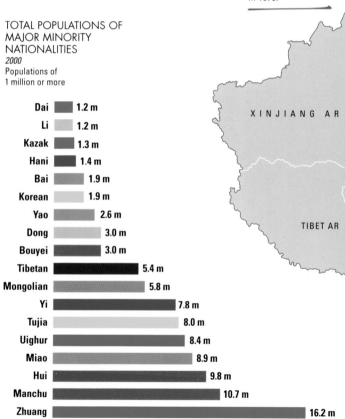

TOTAL POPULATIONS OF
MAJOR MINORITY
NATIONALITIES
2000
Populations of
1 million or more

Dai	1.2 m
Li	1.2 m
Kazak	1.3 m
Hani	1.4 m
Bai	1.9 m
Korean	1.9 m
Yao	2.6 m
Dong	3.0 m
Bouyei	3.0 m
Tibetan	5.4 m
Mongolian	5.8 m
Yi	7.8 m
Tujia	8.0 m
Uighur	8.4 m
Miao	8.9 m
Hui	9.8 m
Manchu	10.7 m
Zhuang	16.2 m

XINJIANG AR

TIBET AR

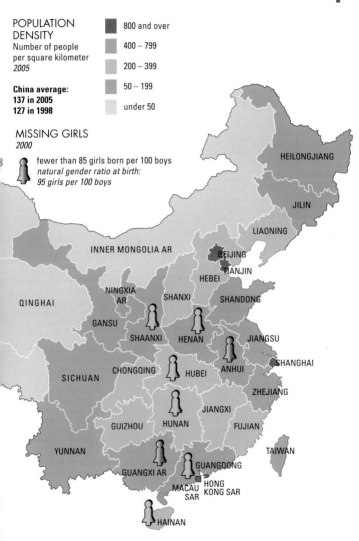

POPULATION
DENSITY
Number of people
per square kilometer
2005

China average:
137 in 2005
127 in 1998

- 800 and over
- 400 – 799
- 200 – 399
- 50 – 199
- under 50

MISSING GIRLS
2000

fewer than 85 girls born per 100 boys
natural gender ratio at birth:
95 girls per 100 boys

HEILONGJIANG

JILIN

LIAONING

INNER MONGOLIA AR

BEIJING

TIANJIN

HEBEI

NINGXIA
AR

SHANXI

SHANDONG

QINGHAI

GANSU

SHAANXI

HENAN

JIANGSU

CHONGQING

HUBEI

ANHUI

SHANGHAI

SICHUAN

ZHEJIANG

GUIZHOU

HUNAN

JIANGXI

FUJIAN

YUNNAN

TAIWAN

GUANGXI AR

GUANGDONG

MACAU
SAR

HONG
KONG SAR

HAINAN

Only five religions are allowed by the government. They are tolerated under restricted conditions which prohibit the use of a religious venue or activities that might harm national or ethnic unity. Chinese indigenous religion fuses Daoism, Buddhism, folk religion and Confucianism, and is widely practised, often alongside other religions.

An estimated 35 million Christians belong to "house churches", not recognized by the government.

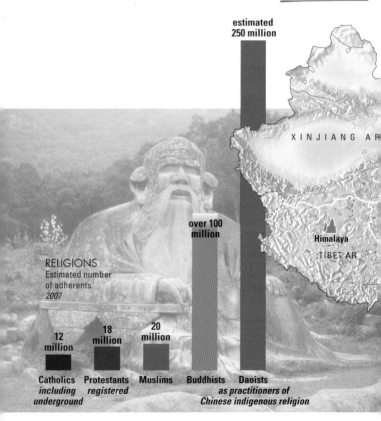

estimated 250 million

XINJIANG AR

over 100 million

Himalaya

TIBET AR

RELIGIONS
Estimated number of adherents
2007

12 million
Catholics
including underground

18 million
Protestants
registered

20 million
Muslims

Buddhists

Daoists
as practitioners of Chinese indigenous religion

RELIGION 中

SACRED SITES
Sites popular with tourists

Major holy mountain

🏔 Buddhist

🏔 Daoist

⭐ Communist Party sacred site

A survey conducted in 2005–06 suggested that nearly a third of adults in China consider themselves religious. →

HEILONGJIANG

JILIN

INNER MONGOLIA AR

LIAONING

⭐ **Beijing:
Tiananmen Square.
People's Republic
of China declared,
1 October 1949**

BEIJING

TIANJIN

Yellow River

Heng-shan

HEBEI

SHANDONG

NINGXIA AR

Wu-tai-shan
SHANXI

Tai-shan

JIANGSU

⭐ **Shanghai:
National Congress
of Communist Party
of China, 1921**

QINGHAI

⭐ **Yan'an:
Headquarters of liberated
areas, 1936–49**

GANSU

HENAN

ANHUI

SHANGHAI

Pu-tuo Is.

SHAANXI

Hua-shan

Song-shan

Jiuhua-shan

ZHEJIANG

Pu-tuo-shan

CHONGQING

HUBEI

SICHUAN

Heng-shan

GUIZHOU

HUNAN

Yangtse River

JIANGXI

FUJIAN

Emei-shan

⭐ **Zunyi City:
Site at which
Mao took over
leadership, 1935**

⭐ **Jinggan Mountain:
first rural revolutionary
base established, 1927**

TAIWAN

GUANGDONG

YUNNAN

GUANGXI AR

⭐ **Guangzhou:
Museum commemorating
the Peasant Movement
Institute, where Mao
taught, 1926**

HAINAN

Part Three
LIVING IN CHINA

THE EMERGING CONSUMER MARKET in China is most evident in the eastern cities, where the availability of branded goods is ubiquitous, and there is pressure, through the political rhetoric of social harmonization, for society to become middle-class. However, there are more people struggling to cope with the "three difficults" (the rising costs of housing, education and health) than there are new rich enjoying the luxuries of globalization.

Education is central to the ethos of parenthood in Chinese society, but there is a division between those whose children are taught in well-equipped modern classrooms, and those whose children attend poorly funded rural schools, with little or no access to technology or a diverse curriculum. Housing is changing, and city prices are rising to unaffordable levels. City developments emphasize modernity and convenience, and older dwellings are pulled down with little or no consultation or compensation. Health is also an expensive commodity, available only to the privileged few.

Most disturbing is the growing problem of food safety. Many staple foods are contaminated by the overuse of fertilizer, the need for which is increased by poor soil care. Meanwhile, the requirement to make agriculture more profitable persuades farmers to produce luxury foods for export, with a reduced attention to sustainable techniques or to a healthy product. The organic food trend has, however, hit the larger cities, where the rich can afford to make some attempt at safeguarding their health.

Universal healthcare has not been available since the Maoist period. The cost of healthcare is a contentious issue in China, although reforms are planned that include universal healthcare insurance.

An estimated 4.5 million people suffer from active TB in China. There is an urgent need for health screening to identify the many millions more carrying the infection, as well as the more recent SARS virus.

More than 7% of household expenditure went on healthcare in 2005.

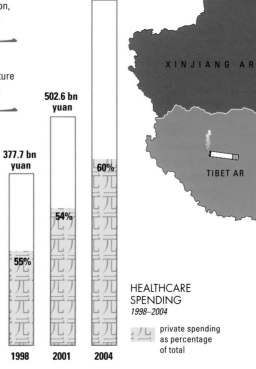

759.0 bn yuan

60%

502.6 bn yuan

54%

377.7 bn yuan

55%

1998   2001   2004

XINJIANG AR

TIBET AR

HEALTHCARE SPENDING
1998–2004

private spending as percentage of total

HEALTHCARE 中

DOCTORS
per 100,000 population
2005

- 200 and over
- 150 – 199
- 100 – 149
- no data

National average: 148

EXPENDITURE ON TOBACCO
As a percentage of expenditure on food by urban dwellers
2005

6% or more

HEILONGJIANG

JILIN

LIAONING

INNER MONGOLIA AR

BEIJING

HEBEI

TIANJIN

NINGXIA AR

SHANXI

QINGHAI

GANSU

SHAANXI

SHANDONG

HENAN

ANHUI

JIANGSU

SICHUAN

CHONGQING

HUBEI

SHANGHAI

ZHEJIANG

JIANGXI

GUIZHOU

HUNAN

FUJIAN

YUNNAN

GUANGXI AR

GUANGDONG

TAIWAN

MACAU SAR

HONG KONG SAR

HAINAN

The number of young, rural, poor children dropping out of school is increasing, despite the government's plan to narrow the inequalities between the cities and the countryside.

Education is expensive but is such a high priority that parents unable to get their children into top-rank Chinese universities are paying to send them overseas. Some parents, unable even to afford university fees in local colleges for their bright offspring, have committed suicide.

In 2005, 2.6 million Chinese were enrolled on internet-based courses, including teacher training, medicine, and engineering.

XINJIANG AR

TIBET AR

OVERSEAS STUDENTS
Number of students studying abroad and number returning
1995–2005

number studying abroad

20,381	1995
38,989	2000
118,515	2005

number returning

5,750	1995
9,121	2000
34,987	2005

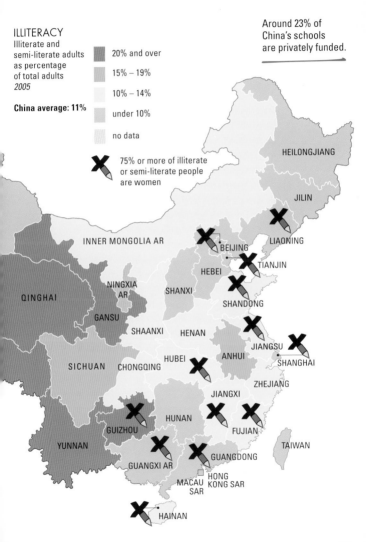

ILLITERACY

Illiterate and
semi-literate adults
as percentage
of total adults
2005

China average: 11%

- 20% and over
- 15% – 19%
- 10% – 14%
- under 10%
- no data

75% or more of illiterate
or semi-literate people
are women

Around 23% of
China's schools
are privately funded.

HEILONGJIANG

JILIN

INNER MONGOLIA AR

LIAONING

BEIJING

TIANJIN

HEBEI

NINGXIA
AR

SHANXI

QINGHAI

GANSU

SHANDONG

SHAANXI

HENAN

JIANGSU

SICHUAN

CHONGQING

HUBEI

ANHUI

SHANGHAI

ZHEJIANG

JIANGXI

GUIZHOU

HUNAN

FUJIAN

YUNNAN

GUANGXI AR

GUANGDONG

TAIWAN

MACAU
SAR

HONG
KONG SAR

HAINAN

People in China are consuming, on average, 1,000 calories a day more than they did in the mid-1960s. Their diet is more varied, but there has been an increase in the number of obese and overweight people.

There are also growing concerns over food safety because of increased use of toxic pesticides – a danger farmers seem unaware of, partly because of poor labeling on imported chemicals.

CHANGING RURAL DIETS
Annual consumption
per person of major foods
in rural households
1985–2005

XINJIANG AR

TIBET AR

	1985	2005
grain	466 kg	209 kg
meat	11 kg	22 kg
fresh vegetables	131 kg	102 kg
liquor	4 liters	8 liters

Ballroom dancing was added to the school curriculum in 2007, in an attempt to combat childhood obesity.

FOOD中

RURAL EXPENDITURE
ON FOOD
As a percentage of
living expenditure
1985–2005

China rural average: 45%
China urban average: 37%

	55% and over
	50% – 54%
	45% – 49%
	40% – 44%
	under 40%
	no data

DINING OUT
As percentage of
expenditure on food
by urban households
2005

20% and over

HEILONGJIANG

JILIN

LIAONING

INNER MONGOLIA AR

BEIJING
HEBEI
TIANJIN

QINGHAI

NINGXIA
AR
SHANXI
SHANDONG

GANSU

SHAANXI
HENAN
JIANGSU

SICHUAN
CHONGQING
HUBEI
ANHUI
SHANGHAI

ZHEJIANG

JIANGXI

GUIZHOU
HUNAN

FUJIAN

TAIWAN

YUNNAN
GUANGXI AR
GUANGDONG

MACAU
SAR
HONG
KONG SAR

HAINAN

The increase in longevity and the success of the small-family policy has meant that the number of working-age people available to support the older age-group is declining. Around three-quarters of China's workforce are not covered by a public pension system. Government plans for a nationwide social insurance system will take time to implement.

The number of retired people is expected to double by 2015 to 200 million.

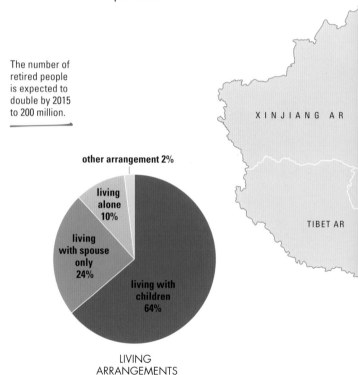

XINJIANG AR

TIBET AR

other arrangement 2%

living alone 10%

living with spouse only 24%

living with children 64%

LIVING ARRANGEMENTS
of people aged 65 and over
2000

WELFARE 中

ELDERLY DEPENDENCY
People aged 65 years and over
as a percentage of those
aged 15–64 years
2005

China average: 12.7%

- 15.0% and over
- 12.5% – 14.9%
- 10.0% – 12.4%
- under 10.0%
- no data

dependency
increased
by a fifth or more
2002–05

China has not fully met its commitment to improve human rights in the lead-up to the Olympic Games. Although all death sentences now have to be reviewed by the highest court in the land, the death penalty can still be given for 68 offences, including non-violent crimes such as fraud, bigamy, and internet hacking.

Nearly 1.5 million people in Beijing have been evicted without adequate compensation during the city's redevelopment for the Olympic Games.

Amnesty estimates the actual number of executions in China in 2006 to have been nearly 8,000 – more than 6 executions per million people.

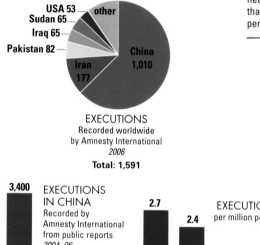

USA 53
Sudan 65
Iraq 65
Pakistan 82
Iran 177
China 1,010
other

EXECUTIONS
Recorded worldwide
by Amnesty International
2006

Total: 1,591

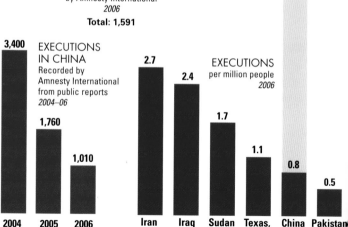

EXECUTIONS IN CHINA
Recorded by Amnesty International from public reports
2004–06

3,400 — 2004
1,760 — 2005
1,010 — 2006

EXECUTIONS
per million people
2006

Iran 2.7
Iraq 2.4
Sudan 1.7
Texas, USA 1.1
China 0.8
Pakistan 0.5

INTELLECTUAL FREEDOM
1957–2007

■ action encouraging freedom

■ action discouraging freedom

1957	Chinese intellectuals, encouraged by Mao to criticize Communist Party, attack its right to govern.
1957–58	Crackdown on intellectuals.
1966–76	Cultural Revolution brings further suppression of intellectuals.
1978	Democracy Wall set up in Beijing, on which citizens are encouraged to paste political tracts.
1979	Wall closed.
1989	Tiananmen Square demonstration for democratic reform is brutally suppressed.
1999	Falun Gong banned under law against cults.
2000	Party membership opened up to entrepreneurs, intellectuals and scientists.
2002	Government promotes the internet because of its economic benefits, but exercises control over information.
2004	Human rights guaranteed and private property protected under amendment to state constitution.
2004	Discrimination against people with HIV/AIDS is banned.
2005	Government publishes first White Paper on democracy, but little is proposed in the way of political reforms.
2006	Confucianism, previously banned, encouraged by government.
2006	Harmonious Socialist Society proposed by government to address inequalities.

By 2020 China is expected to be a predominantly urban society, with up to 60 percent of its population living in towns and cities.

Up to 200 million people have left rural areas in recent years to seek work in cities. In 2005, 36% of rural incomes derived from remittances sent home by migrant workers in the cities.

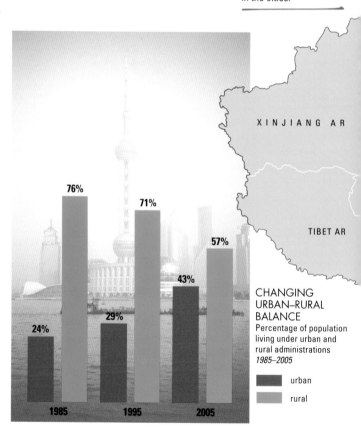

XINJIANG AR

TIBET AR

76%

71%

57%

43%

29%

24%

1985

1995

2005

CHANGING URBAN–RURAL BALANCE
Percentage of population living under urban and rural administrations
1985–2005

■ urban
■ rural

URBAN POPULATION
As a percentage of
total population
of province
2005

- 75% and over
- 50% – 74%
- 40% – 49%
- under 40%
- no data

URBAN GROWTH
Percentage increase in
urban population
2000–05

- 10% and over
- 5% – 9%

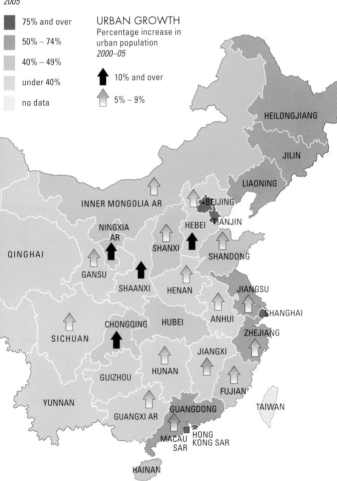

HEILONGJIANG

JILIN

LIAONING

INNER MONGOLIA AR

BEIJING

TIANJIN

NINGXIA
AR

HEBEI

QINGHAI

SHANXI

SHANDONG

GANSU

SHAANXI

HENAN

JIANGSU

SHANGHAI

CHONGQING

HUBEI

ANHUI

ZHEJIANG

SICHUAN

JIANGXI

HUNAN

GUIZHOU

FUJIAN

YUNNAN

GUANGDONG

TAIWAN

GUANGXI AR

MACAU
SAR

HONG
KONG SAR

HAINAN

China's car production reached 5.2 million vehicles in 2006, compared with US production of 4.4 million. In Beijing alone, 1,000 new cars are produced every day.

Domestic flights are becoming much more widely used, the number of passengers having doubled between 2000 and 2005.

Having constructed the highest railroad in the world, China plans to build the highest airport – 4,334 metres (14,220 feet) above sea-level at Ngari, Tibet.

AIR TRAVEL
Number of international and domestic passengers
1990–2005

1990
17 million

1995
51 million

2000
67 million

2005
138 million

XINJIANG AR

TIBET AR

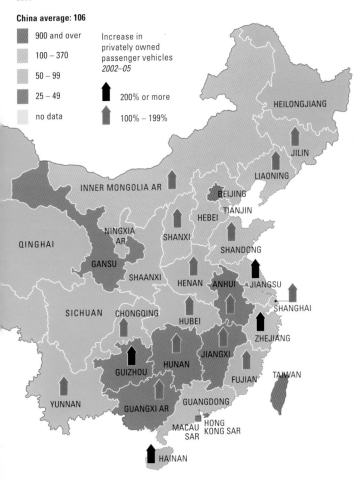

TRANSPORT 中

PRIVATE TRANSPORT
Number of passenger
vehicles owned
per 10,000 people
2005

China average: 106

■	900 and over
▨	100 – 370
▢	50 – 99
■	25 – 49
▨	no data

Increase in
privately owned
passenger vehicles
2002–05

⬆ 200% or more

⬆ 100% – 199%

HEILONGJIANG

JILIN

LIAONING

INNER MONGOLIA AR

BEIJING

HEBEI

TIANJIN

NINGXIA AR

SHANXI

SHANDONG

QINGHAI

GANSU

SHAANXI

HENAN

ANHUI

JIANGSU

SICHUAN

CHONGQING

HUBEI

SHANGHAI

ZHEJIANG

GUIZHOU

HUNAN

JIANGXI

YUNNAN

GUANGXI AR

GUANGDONG

FUJIAN

TAIWAN

MACAU SAR

HONG KONG SAR

HAINAN

37

China's consumption of fossil fuels rose by 9.3 percent in 2006, about eight times the US increase of 1.2 percent. China is overtaking the USA as the world's largest carbon emitter, although emissions per person are far below those in the West.

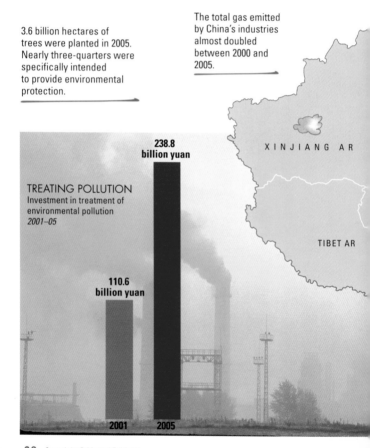

3.6 billion hectares of trees were planted in 2005. Nearly three-quarters were specifically intended to provide environmental protection.

The total gas emitted by China's industries almost doubled between 2000 and 2005.

XINJIANG AR

238.8 billion yuan

TREATING POLLUTION
Investment in treatment of environmental pollution
2001–05

TIBET AR

110.6 billion yuan

2001   2005

POLLUTION 中

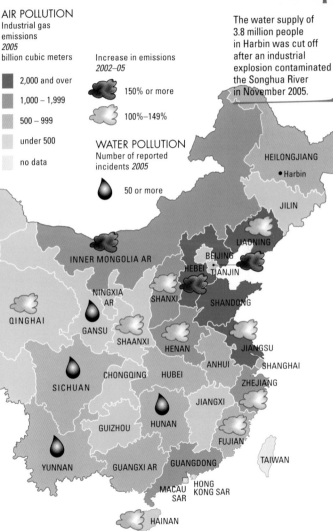

AIR POLLUTION
Industrial gas emissions
2005
billion cubic meters

- 2,000 and over
- 1,000 – 1,999
- 500 – 999
- under 500
- no data

Increase in emissions
2002–05

- 150% or more
- 100%–149%

WATER POLLUTION
Number of reported incidents *2005*

- 50 or more

The water supply of 3.8 million people in Harbin was cut off after an industrial explosion contaminated the Songhua River in November 2005.

HEILONGJIANG

• Harbin

JILIN

LIAONING

INNER MONGOLIA AR

BEIJING

HEBEI

TIANJIN

SHANXI

SHANDONG

NINGXIA AR

QINGHAI

GANSU

SHAANXI

HENAN

JIANGSU

CHONGQING

HUBEI

ANHUI

SHANGHAI

ZHEJIANG

SICHUAN

JIANGXI

HUNAN

GUIZHOU

FUJIAN

YUNNAN

GUANGXI AR

GUANGDONG

TAIWAN

MACAU SAR

HONG KONG SAR

HAINAN

A chronic lack of water in northeast China is made worse by pollution of both rivers and groundwater. The result is that more than 320 million rural residents do not have clean drinking water, and 400 of China's largest cities are facing a water shortage. An ambitious scheme to divert water from south to north will meet but a fraction of the need, and the water will be subject to pollution in the process. For the Olympic Games, the capital is planning a water transfer from Shanxi Province.

A 2006 study predicted that the Yangtze water would be unusable in 10 years unless the discharge of untreated wastewater was stopped.

XINJIANG AR

TIBET AR

WATER USE
by sector
2005

domestic 12%
industry 23%
agriculture 65%

Total used: 563 cubic kilometers

WATER 中

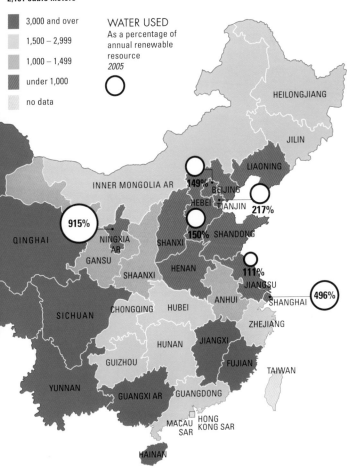

WATER AVAILABILITY
per person per year
2005 cubic meters

**China average:
2,151 cubic meters**

- 3,000 and over
- 1,500 – 2,999
- 1,000 – 1,499
- under 1,000
- no data

WATER USED
As a percentage of
annual renewable
resource
2005

HEILONGJIANG

JILIN

LIAONING

INNER MONGOLIA AR

149%

BEIJING

HEBEI

TIANJIN

217%

915%

NINGXIA
AR

GANSU

SHANXI

150% SHANDONG

QINGHAI

SHAANXI

HENAN

111%

JIANGSU

SHANGHAI

496%

SICHUAN

CHONGQING

HUBEI

ANHUI

ZHEJIANG

HUNAN

JIANGXI

GUIZHOU

FUJIAN

TAIWAN

YUNNAN

GUANGXI AR

GUANGDONG

MACAU
SAR

HONG
KONG SAR

HAINAN

Part Five
ECONOMY

SINCE 1978, WHEN MAO'S SUCCESSOR Deng Xiaoping urged those with entrepreneurial spirit to get rich, China's economic rise has been premised on its status as the world's factory, and industrial enterprises have grown to meet global demand. Many people have indeed got rich, and the standard of living for many more has improved, while others have failed to reap the rewards of a growing market economy. The result is a mixture of world-class companies and lumbering state enterprises, and an ongoing reliance on exports to the developed world. Chinese factories are pumping goods into other economies, and all but the most unusual, localized, or luxurious consumer goods are now marked "Made in China".

China's growth is not without benefits beyond its own borders, however. The country is not self-sufficient, and requires the import of oil and gas from Africa, Australia, and South America. In particular, China is working to make Africa not only a second source of raw materials, but a useful site for offshore production of Chinese goods. Government policy is seeking to change China's image from that of a source of cheap labor to a supplier of creative and innovative employees, reliant on education and skill, and with a keen eye on the benefits of intellectual property. This is a natural progression in a global economy, and a warning to rich nations such as the USA to look after their intellectual and creative capital before a brain drain to China strips their major assets.

Since 1990, China's economy has grown by an average of 10 percent a year. In order to create a more balanced economy across China's provinces, the government has been encouraging enterprises to move inland and to "open up to the West". But although labor is cheaper in these areas, there are fewer skilled laborers, and transportation costs are higher.

In 2006, China was the fourth-largest economy in the world, but it is expected to overtake Germany and Japan to become the second-largest economy behind the USA within 10 or 15 years.

XINJIANG AR

TIBET AR

NATIONAL GDP
1985–2005
billion yuan

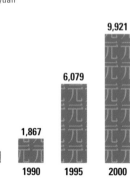

902	1,867	6,079	9,921	18,308
1985	1990	1995	2000	2005

ECONOMIC GROWTH 中

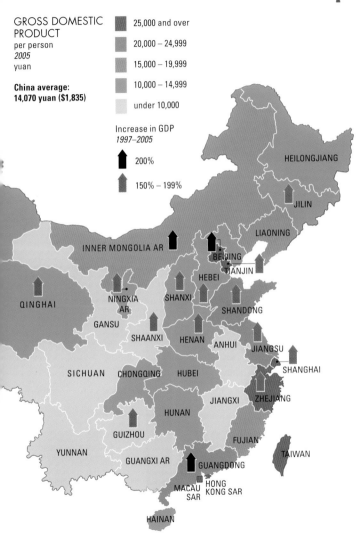

GROSS DOMESTIC
PRODUCT
per person
2005
yuan

**China average:
14,070 yuan ($1,835)**

25,000 and over

20,000 – 24,999

15,000 – 19,999

10,000 – 14,999

under 10,000

Increase in GDP
1997–2005

200%

150% – 199%

HEILONGJIANG

JILIN

LIAONING

INNER MONGOLIA AR

BEIJING

TIANJIN

HEBEI

QINGHAI

NINGXIA
AR

SHANXI

SHANDONG

GANSU

SHAANXI

HENAN

ANHUI

JIANGSU

SICHUAN

CHONGQING

HUBEI

SHANGHAI

JIANGXI

ZHEJIANG

HUNAN

GUIZHOU

FUJIAN

YUNNAN

GUANGXI AR

GUANGDONG

TAIWAN

MACAU
SAR

HONG
KONG SAR

HAINAN

Low wages, high productivity and better skills contribute to China's international competitiveness. There is a shift in employment from agriculture to services, such as tourism, hospitality, banking and finance, as China develops a more diverse economy.

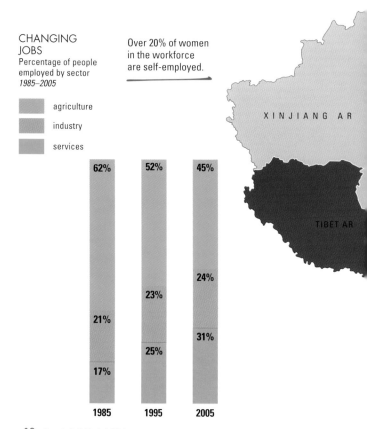

CHANGING
JOBS
Percentage of people
employed by sector
1985–2005

Over 20% of women
in the workforce
are self-employed.

agriculture
industry
services

	1985	1995	2005
agriculture	62%	52%	45%
industry	21%	23%	24%
services	17%	25%	31%

XINJIANG AR

TIBET AR

EMPLOYMENT 中

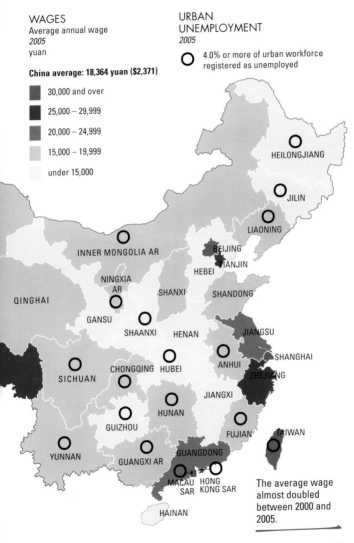

WAGES
Average annual wage
2005
yuan

China average: 18,364 yuan ($2,371)

- 30,000 and over
- 25,000 – 29,999
- 20,000 – 24,999
- 15,000 – 19,999
- under 15,000

URBAN
UNEMPLOYMENT
2005

○ 4.0% or more of urban workforce registered as unemployed

HEILONGJIANG

JILIN

LIAONING

INNER MONGOLIA AR

BEIJING

TIANJIN

HEBEI

NINGXIA AR

SHANXI

SHANDONG

QINGHAI

GANSU

SHAANXI

HENAN

JIANGSU

SHANGHAI

SICHUAN

CHONGQING

HUBEI

ANHUI

ZHEJIANG

JIANGXI

HUNAN

GUIZHOU

FUJIAN

TAIWAN

YUNNAN

GUANGXI AR

GUANGDONG

MACAU SAR

HONG KONG SAR

HAINAN

The average wage almost doubled between 2000 and 2005.

49

The government aims to reduce inequalities between rich and poor, cities and countryside, and coastal and western regions, to create a "harmonious socialist society" by 2020. This strategy indicates both real intent and the need to cover up the pain of development experienced by many workers and farmers, whether they stay on the land or migrate to other provinces for employment.

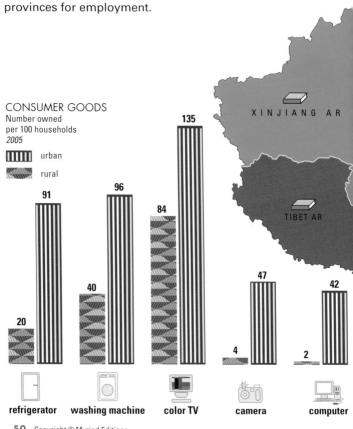

CONSUMER GOODS
Number owned
per 100 households
2005

▓▓▓ urban
▨▨▨ rural

	refrigerator	washing machine	color TV	camera	computer
urban	91	96	135	47	42
rural	20	40	84	4	2

XINJIANG AR

TIBET AR

INEQUALITY 中

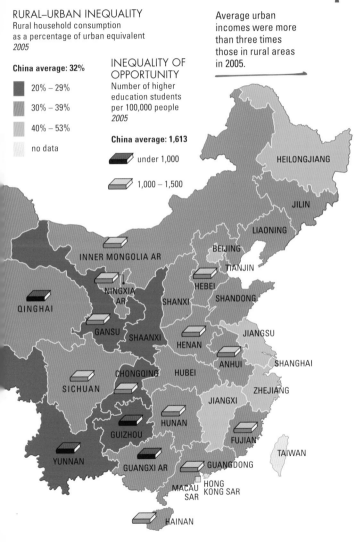

RURAL–URBAN INEQUALITY
Rural household consumption
as a percentage of urban equivalent
2005

China average: 32%

- 20% – 29%
- 30% – 39%
- 40% – 53%
- no data

INEQUALITY OF OPPORTUNITY
Number of higher
education students
per 100,000 people
2005

China average: 1,613

- under 1,000
- 1,000 – 1,500

Average urban
incomes were more
than three times
those in rural areas
in 2005.

HEILONGJIANG

JILIN

LIAONING

BEIJING

TIANJIN

INNER MONGOLIA AR

NINGXIA AR

HEBEI

SHANDONG

SHANXI

QINGHAI

GANSU

SHAANXI

HENAN

JIANGSU

ANHUI

SHANGHAI

CHONGQING

HUBEI

SICHUAN

ZHEJIANG

JIANGXI

GUIZHOU

HUNAN

FUJIAN

YUNNAN

GUANGXI AR

GUANGDONG

TAIWAN

MACAU SAR

HONG KONG SAR

HAINAN

China ranked fourth in the world as a destination for overseas visitors in 2006. Tourism will continue to grow, with the Beijing Olympics in 2008 and the World Expo in Shanghai in 2010 attracting both international and domestic tourists.

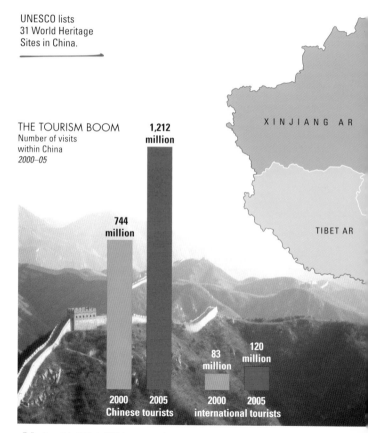

UNESCO lists 31 World Heritage Sites in China.

THE TOURISM BOOM
Number of visits within China
2000–05

1,212 million

744 million

120 million

83 million

2000	2005	2000	2005
Chinese tourists		international tourists	

XINJIANG AR

TIBET AR

TOURISM 中

FOREIGN TOURISTS
Number of tourist visits
by foreigners
2005

- 2 million and over
- 1 million – 2 million
- 500,000 – 999,999
- 250,000 – 499,999
- under 250,000
- no data

number of tourists
more than quadrupled
1995–2005

INCOME FROM TOURISM
Foreign exchange earnings
from tourism in 2005

$ = US$ 1 billion

HEILONGJIANG

JILIN

LIAONING

INNER MONGOLIA AR

BEIJING

$$$

HEBEI TIANJIN

SHANXI

SHANDONG

NINGXIA AR

QINGHAI

GANSU

SHAANXI

HENAN

ANHUI

JIANGSU

$$

6.1 million people
visited Shanghai
in 2006, including
1.1 million from
Japan.

$$$

SHANGHAI

SICHUAN CHONGQING

HUBEI

ZHEJIANG

$

JIANGXI

GUIZHOU HUNAN

$

FUJIAN

TAIWAN

YUNNAN

GUANGXI AR GUANGDONG

$$$$$$

MACAU
SAR

HONG
KONG SAR

HAINAN

55

China is set to overtake the USA and become the largest importer, as its booming economy draws in goods and raw materials, along with foreign investment. The value of China's exports equals or exceeds that of its imports, and its overseas investments and aid are turning it into an economic superpower.

In 2007 Wal-Mart concluded a deal giving it an interest in 100 stores in 34 cities. Carrefour had 90 hypermarkets.

TRADE BOOM
Value of total
imports and exports
1985–2005
US$ billion

$$\text{imports}$$

$$\text{exports}$$

XINJIANG AR

TIBET AR

$762
$660
$149
$132
$42
$27

1985   1995   2005

TRADE AND INVESTMENT 中

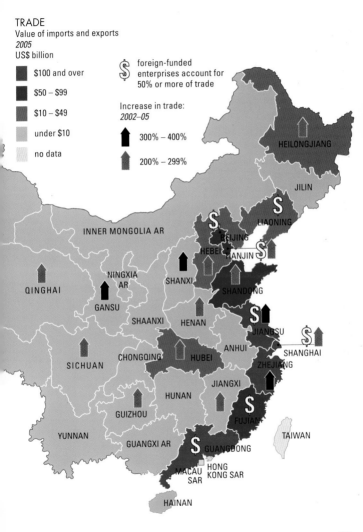

TRADE
Value of imports and exports
2005
US$ billion

- $100 and over
- $50 – $99
- $10 – $49
- under $10
- no data

$ foreign-funded enterprises account for 50% or more of trade

Increase in trade:
2002–05

- 300% – 400%
- 200% – 299%

HEILONGJIANG

JILIN

INNER MONGOLIA AR

LIAONING

BEIJING

HEBEI

TIANJIN

NINGXIA AR

SHANXI

SHANDONG

QINGHAI

GANSU

SHAANXI

HENAN

JIANGSU

ANHUI

SHANGHAI

SICHUAN

CHONGQING

HUBEI

ZHEJIANG

JIANGXI

HUNAN

GUIZHOU

FUJIAN

YUNNAN

GUANGXI AR

GUANGDONG

MACAU SAR

HONG KONG SAR

TAIWAN

HAINAN

57

Despite China's military modernization, it is unlikely that it will become a military competitor of the USA in the near future. Taiwan, however, remains a potential point of conflict. China claims sovereignty over the island, where the Nationalists set up a separate government at the end of the civil war in 1949. It promotes peaceful reunification, but retains the right to use force if necessary. The USA is committed to supplying Taiwan with defensive weapons, but at the same time opposes Taiwan's independence, and denies it the statehood necessary for it to join international organizations.

Although the tensions and military build up persist on both sides of the Taiwan Strait, Taiwanese investment in mainland China exceeds US$50 billion, and more than a million Taiwanese live there.

China's military budget grew by 17.8% in 2007, the largest increase for more than a decade. Yet it is less than one-half the combined budget of the UK, France, Germany and Italy.

CHINA'S DEFENSE SPENDING
1985–2005

247 billion yuan
2005

64 billion yuan
1995

19 billion yuan
1985

CHINA VS TAIWAN
2006

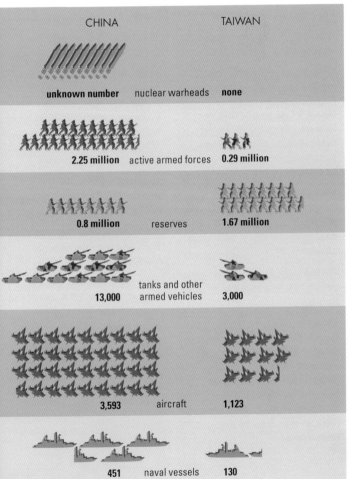

	CHINA		TAIWAN
	unknown number	nuclear warheads	**none**
	2.25 million	active armed forces	**0.29 million**
	0.8 million	reserves	**1.67 million**
	13,000	tanks and other armed vehicles	**3,000**
	3,593	aircraft	**1,123**
	451	naval vessels	**130**

Source: IISS, Military Balance 2006

POPULATION

	Total population 2005 millions	Number of girls born per 100 boys 2000	Urban as % of total 2005	Minority nationalities as % of population 2000	Illiterate adults as % of population 2005
Total	**1,306.3**	**86**	**43%**	**8%**	**11%**
Anhui	61	84	36%	1%	19%
Beijing	15	89	84%	4%	4%
Chongqing	28	88	45%	7%	12%
Fujian	35	87	47%	2%	13%
Gansu	26	88	30%	9%	21%
Guangdong	92	85	61%	2%	6%
Guangxi	47	88	34%	38%	9%
Guizhou	37	89	27%	38%	21%
Hainan	8	82	45%	17%	10%
Hebei	68	90	38%	4%	7%
Heilongjiang	38	90	53%	5%	6%
Henan	94	85	31%	1%	10%
Hubei	57	90	43%	4%	12%
Hunan	63	85	37%	10%	9%
Inner Mongolia	24	91	47%	21%	11%
Jiangsu	75	87	50%	0%	10%
Jiangxi	43	80	37%	0%	11%
Jilin	27	93	53%	9%	6%
Liaoning	42	92	59%	16%	5%
Ningxia	6	91	42%	35%	19%
Qinghai	5	91	39%	46%	24%
Shaanxi	37	88	37%	1%	10%
Shandong	92	92	45%	1%	12%
Shanghai	18	89	89%	1%	5%
Shanxi	34	90	42%	0%	6%
Sichuan	82	93	33%	5%	17%
Tianjin	10	94	75%	3%	5%
Tibet	3	93	27%	94%	45%
Xinjiang	20	95	37%	59%	8%
Yunnan	44	89	30%	33%	20%
Zhejiang	49	87	56%	1%	12%
Hong Kong SAR	7				7%
Macau SAR	0.5				5%
Taiwan	23				4%

Unless otherwise specified, all data on the maps are from the *China Statistical Yearbook*.

DATA TABLE 中

ECONOMY		ENVIRONMENT		
Gross regional product per person *2005* yuan	Average wage *2005* yuan	Water available per person per year *2005* cubic meters	Industrial gas emissions *2005* billion cubic meters	Passenger vehicles owned per 10,000 people *2005*
14,002	**18,364**	**2,152**	**26,899**	**106**
8,783	15,334	1,179	696	38
44,774	34,191	151	353	914
10,974	16,630	1,827	366	55
18,583	17,146	3,976	627	83
7,456	14,939	1,042	425	27
24,327	23,959	1,906	1,345	199
8,746	15,461	3,704	834	45
5,306	14,344	2,244	385	44
10,804	14,417	3,722	91	56
14,737	14,707	197	2,652	142
14,428	14,458	1,954	526	86
11,287	14,282	597	1,550	65
11,419	14,419	1,641	940	52
10,293	15,659	2,650	601	47
16,327	15,985	1,917	1,207	116
24,489	20,957	627	2,020	126
9,410	13,688	3,513	438	27
13,329	14,409	2,067	494	108
18,974	17,331	896	2,090	112
10,169	17,211	144	284	78
10,006	19,084	16,177	137	58
9,881	14,796	1,323	492	70
20,023	16,614	451	2,413	106
51,486	34,345	138	848	230
12,458	15,645	252	1,514	128
8,993	15,826	3,570	814	81
35,452	25,271	102	460	370
9,069	28,950	161,171	1	54
12,956	15,558	4,809	449	94
7,804	16,140	4,162	544	89
27,435	25,896	2,077	1,303	210
197,857	117,755 *			779†
178,000	69,180 *			3,229†
118,015	122,711 *			8,720†

* converted into yuan   † all vehicles
from local currency

中 INDEX

Additional research:
Tina Schilbach, UTS China Research Centre.

By the same authors:

The State of China Atlas
Mapping the World's Fastest Growing Economy
2nd edition

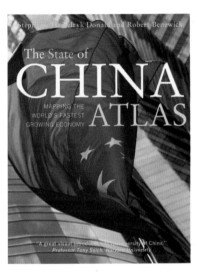

"Simply indispensable for teaching...clear, comprehensive, and focused on the most crucial issues facing the country. It has provoked some of the best discussions we have had in class."
Professor Marc Blecher, Oberlin College, Ohio

"It is still difficult to get reliable hard data from China. [This] fills a vital gap, ranging from population, economic growth and political leadership through to social and environmental development."
John Gittings, former Asia editor, *Guardian*

"Colorful, attention-grabbing graphics along with solid data and handy up-to-date information on a wide variety of topics. Keep a copy close-by."
China Information

Available from the University of California Press

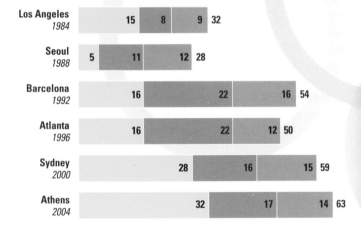

MEDAL TALLY
Olympic medals
won by China
1984–2004

gold   silver   bronze   total

Los Angeles
1984
15   8   9   32

Seoul
1988
5   11   12   28

Barcelona
1992
16   22   16   54

Atlanta
1996
16   22   12   50

Sydney
2000
28   16   15   59

Athens
2004
32   17   14   63